K. CONNORS

Into the Unknown

A Journey through Deep Questions and Theories

Copyright © 2024 by K. Connors

All rights reserved. No part of this publication may be reproduced, stored or transmitted in any form or by any means, electronic, mechanical, photocopying, recording, scanning, or otherwise without written permission from the publisher. It is illegal to copy this book, post it to a website, or distribute it by any other means without permission.

First edition

This book was professionally typeset on Reedsy.
Find out more at reedsy.com

Contents

Introduction:	1
Chapter 1: Unraveling Reality	3
Chapter 2: Navigating Time	6
Chapter 3: The Mind's Dilemma	9
Chapter 4: Multiverse Mayhem	12
Chapter 5: The Fermi Paradox	15
Chapter 6: The Meaning of Life	18
Chapter 7: Infinity and Finite Minds	21
Chapter 8: The Simulation Hypothesis	24
Chapter 9: The Uncertainty Principle	27
Chapter 10: Existentialism and Absurdism	30
Chapter 11: The Chicken or the Egg	33
Conclusion:	36

Introduction:

Welcome, explorers of the curious and the profound, to a journey through the labyrinth of human thought and inquiry. In this book, we embark on a quest to unravel the mysteries of existence, to ponder the deepest questions that have captivated the minds of philosophers, scientists, and dreamers for centuries.

From the nature of reality itself to the origins of life and consciousness, we delve into the heart of the unknown, armed with nothing but our curiosity and a thirst for understanding. Along the way, we encounter mind-bending paradoxes, cosmic conundrums, and existential quandaries that challenge our very understanding of the universe.

Our journey begins with an exploration of The Chicken or the Egg Question, a timeless puzzle that invites us to ponder the origins of life and the nature of causality. From there, we traverse the realms of quantum mechanics, grappling with the Uncertainty Principle and the strange and wondrous world of subatomic particles.

But our quest doesn't stop there. We venture into the realms of existentialism and absurdism, confronting the existential angst and philosophical musings that define the human experience. We ponder the meaning of life, the nature of consciousness, and the mysteries of existence itself.

Along the way, we encounter concepts both familiar and strange, from the

Simulation Hypothesis to the Fermi Paradox, each offering a tantalizing glimpse into the vast expanse of the unknown. And as we journey deeper into the cosmic unknown, we find ourselves confronted with questions that challenge our most deeply-held beliefs and assumptions about the nature of reality.

But fear not, for this is not merely a journey of intellectual inquiry, but also a journey of wonder, curiosity, and discovery. Whether we find ourselves pondering the mysteries of the universe or simply marveling at the sheer audacity of existence, one thing's for certain: the journey is its own reward.

So, join me as we embark on this odyssey through the realms of thought and inquiry, a journey that promises to challenge, enlighten, and inspire. Together, let's explore the depths of the human mind and the mysteries of the cosmos, one question at a time.

Chapter 1: Unraveling Reality

Welcome to the wild, wacky world of reality—or at least, what we think is reality. Grab your favorite thinking cap and buckle up because we're about to take a joyride through the cosmos of consciousness and existence.

Now, picture this: you're sipping your morning coffee, pondering life's deepest mysteries. You glance around and wonder, "Am I really here, or am I just a character in some cosmic video game, waiting for someone to hit pause?" It's the kind of thought that might make your brain do a double take and reach for the "reboot" button.

But hey, before you start questioning the legitimacy of your existence, let's dive into the rabbit hole of the nature of reality itself. You see, reality is like that elusive friend who always keeps you guessing. Is it solid, tangible, and unquestionably real? Or is it more like a hologram, shimmering and shifting with every glance?

One of the juiciest theories to nibble on is the idea that we're all just players in the ultimate simulation. Picture it: a cosmic computer program, running on some super-advanced alien MacBook, where every quirk, twist, and turn of our lives is just a line of code. It's like "The Sims," but with way higher stakes.

Now, I know what you're thinking: "But wait, wouldn't there be glitches in the matrix if this were true?" Ah friend, that's where things get interesting.

Some theorists argue that the glitches—the déjà vu moments, the inexplicable coincidences—are just the universe's way of winking at us, reminding us that reality isn't as solid as it seems.

Of course, not everyone's convinced. There are skeptics aplenty who scoff at the idea of living in a digital dreamland. They'll tell you, "Come on, get real! This isn't 'The Matrix'—we're flesh-and-blood beings living in a physical world." And sure, they have a point. After all, you can't exactly press Ctrl+Alt+Delete to escape your problems.

But here's the thing: whether we're living in a simulation or not, the mere fact that we're asking these questions speaks volumes about the human spirit. We're curious creatures, always poking and prodding at the fabric of reality, trying to unravel its secrets like cosmic detectives.

So, as we journey deeper into the rabbit hole of reality, keep an open mind and a firm grip on your sense of wonder. Who knows? Maybe by the end of this ride, we'll have unlocked the ultimate cheat code to existence—or at least gained a few extra lives in the game of life.

As we near the end of our expedition through the complications of reality, it's only fitting to pause and reflect on what we've uncovered. We've danced with the idea of living in a simulation, pondered the glitches in the matrix, and marveled at the sheer audacity of existence itself.

But here's the kicker: whether our reality is a finely-tuned simulation or an organic masterpiece, one thing remains abundantly clear—we're in this together. From the humblest earthworm to the mightiest supernova, we're all players on the cosmic stage, each with our own role to play.

So, what's the takeaway from all this existential exploration? Well, for starters, it's a reminder to embrace the uncertainty of it all. Life's a wild, unpredictable ride, and trying to pin it down with neat little labels and categories is like

CHAPTER 1: UNRAVELING REALITY

trying to catch a cloud with a butterfly net.

But amidst the chaos and confusion, there's also beauty to be found. The very fact that we're here, pondering the mysteries of the universe, is a testament to the resilience of the human spirit. We're the dreamers, the seekers, the ones who refuse to take reality at face value.

So, as you go about your day, remember to look up at the stars and marvel at the sheer improbability of it all. Whether we're living in a simulation or not, the journey is what matters—the twists, the turns, the moments of pure, unadulterated awe.

And who knows? Maybe one day, we'll unlock the secrets of the universe and discover the true nature of reality. Or maybe we'll just keep asking questions, chasing after shadows, and reveling in the sheer absurdity of it all.

Either way, one thing's for sure: reality may be a mystery wrapped in an enigma, but it's a mystery worth exploring. So, grab your magnifying glass and your sense of wonder, and let's continue our journey through the cosmos of consciousness.

Who knows what we'll find? Maybe, just maybe, we'll stumble upon the ultimate truth—the answer to the age-old question of what it means to be alive. But until then, let's keep asking, keep searching, and keep marveling at the grandeur of it all. After all, isn't that what makes life worth living?

Chapter 2: Navigating Time

Alright, buckle up your DeLoreans and adjust your flux capacitors because we're about to dive headfirst into the swirling vortex of time travel paradoxes. Yeah, you heard me right—we're about to tackle the big questions like, "Can you change the past without messing up the present?" and "If you go back in time and step on a butterfly, will it really cause a hurricane?"

Now, time travel might sound like something straight out of a sci-fi flick, but trust me, it's a real mind-bender. Just think about it: the past, the present, the future—all tangled up in a spaghetti-like mess that even Nonna would have trouble untangling.

But before we get too tangled up ourselves, let's start with the basics. Picture time as a big ol' river, flowing from the past to the future with a steady, unstoppable current. Now, imagine you've got yourself a shiny new time machine, courtesy of Doc Brown or some other mad scientist. You hop in, hit the gas, and suddenly, you're surfing the waves of time like a pro.

But here's where things get tricky: if you can travel back in time and change the past, what does that mean for the present? Are we talking about alternate timelines, parallel universes, or just a really big headache?

Some folks argue that changing the past is a big no-no, like trying to unscramble an egg or un-bake a cake. According to them, messing with the

CHAPTER 2: NAVIGATING TIME

timeline could cause all sorts of paradoxes—like the classic "grandfather paradox," where you go back in time and accidentally bump off your own grandpa before your parents are born. Talk about awkward family reunions!

But hey, not everyone's convinced. Some daring adventurers think that time is more like a big ol' ball of wibbly-wobbly, timey-wimey stuff, to quote a certain Time Lord. According to them, changing the past might not be such a bad idea after all—heck, it might even be necessary to set things right.

Of course, there's only one way to find out: fire up the time machine and see what happens. Just be sure to pack a spare pair of socks and a healthy dose of skepticism—you never know what kind of temporal shenanigans you might run into out there.

So, as we strap on our time-traveling boots and prepare to leap headfirst into the unknown, remember this: time travel might be a wild ride, but it's also a heck of a lot of fun. So buckle up, hold on tight, and get ready to make some history—or rewrite it, if you're feeling particularly adventurous.

As we approach the end of this whirlwind tour through the timey-wimey realm of time travel, it's only fair to pause and ponder the implications of our temporal escapades. We've danced with paradoxes, flirted with alternate timelines, and maybe even stepped on a few metaphorical butterflies along the way.

But amidst all the chaos and confusion, one thing remains abundantly clear: time is a slippery little devil, always slipping through our fingers like sand in an hourglass. Whether we're hurtling through the past, present, or future, one thing's for sure—we're along for the ride, whether we like it or not.

So, what's the moral of the story? Well, for starters, it's a reminder to tread lightly when playing with the fabric of time. Like a delicate soufflé or a Jenga tower on the brink of collapse, the timeline is a fragile thing, easily upset by even the slightest disturbance.

But hey, that doesn't mean we should shy away from the thrill of temporal exploration. After all, where's the fun in playing it safe? Sometimes, you've gotta take a leap of faith, throw caution to the wind, and see where the tides of time take you.

And who knows? Maybe, just maybe, we'll unlock the secrets of the universe, unravel the mysteries of time itself, and emerge victorious on the other side. Or maybe we'll just end up with a killer case of jet lag and a newfound appreciation for the present moment. Either way, it's bound to be one heck of a ride.

So, as we say farewell to the swirling currents of time and space, let's raise a glass to the intrepid explorers who dare to defy the laws of physics and venture where few have gone before. Here's to the dreamers, the schemers, and the mad scientists with a twinkle in their eye and a time machine in their garage.

And who knows? Maybe, just maybe, the next time you look up at the stars, you'll catch a glimpse of a shooting star—or maybe even a time-traveling DeLorean, streaking across the sky like a comet on a collision course with destiny.

But until that day comes, remember this: whether you're traveling through time or just muddling through the present, the journey is what matters most. So strap in, hold on tight, and get ready for the adventure of a lifetime. After all, time waits for no one—so why should we?

Chapter 3: The Mind's Dilemma

Alright, it's time to dive headfirst into the murky waters of consciousness and free will. Get ready to stretch those brain muscles and ponder some of the juiciest questions known to humankind. Like, do we really have free will, or are we just puppets dancing on the strings of fate? And what's the deal with consciousness, anyway? Is it just a cosmic joke, or is there more to it than meets the eye?

Let's start with free will—or as I like to call it, the eternal dilemma of choice. Picture yourself standing at a crossroads, faced with a seemingly endless array of paths stretching out before you. Do you go left, right, or straight ahead? Or do you just throw your hands up in the air and let fate take the wheel?

Now, some folks will tell you that free will is just an illusion, a trick of the mind designed to make us feel like we're in control. According to them, every choice we make is predetermined by a complex web of biological, psychological, and environmental factors, like some cosmic game of dominoes.

But hey, don't lose hope just yet. Others argue that free will is the real deal, a fundamental aspect of human nature that sets us apart from the rest of the animal kingdom. According to them, we're not just passive observers in the grand symphony of life—we're active participants, wielding the power to shape our own destiny.

Of course, the truth probably lies somewhere in the middle. Maybe we're not completely free to choose our actions, but that doesn't mean we're powerless either. Like a surfer riding the waves of fate, we might not be able to control the ocean, but we can certainly learn to ride the tide.

Now, let's talk about everyone's favorite head-scratcher: consciousness. What is it, where does it come from, and why do we even bother with it? It's like trying to catch a cloud with a butterfly net—elusive, enigmatic, and downright confounding.

Some folks think that consciousness is just a byproduct of our brains firing on all cylinders, like a computer running a particularly complex program. According to them, it's nothing more than a trick of the mind, a cosmic accident that gives us the illusion of self-awareness.

But hold on just a minute. Others argue that consciousness is the real deal, a fundamental aspect of reality that transcends mere neurons and synapses. According to them, it's the spark that ignites the flame of existence, the light that illuminates the darkest corners of the universe.

So, what's the verdict? Well, that's for you to decide friend. Whether you believe in free will, consciousness, or just the power of a good cup of coffee, one thing's for sure: the mind is a mighty mysterious place, full of twists, turns, and tantalizing questions just waiting to be answered.

So, grab your thinking cap and your favorite beverage, and let's embark on a journey through the maze of the mind. Who knows what we'll find? Maybe a few answers, maybe a few more questions, or maybe just a newfound appreciation for the sheer absurdity of it all.

As we near the end of our expedition into the labyrinthine depths of consciousness and free will, it's time to take stock of what we've uncovered. We've danced with the tantalizing concept of free will, pondered the intricacies of

CHAPTER 3: THE MIND'S DILEMMA

choice and destiny, and tiptoed through the murky waters of consciousness, trying to catch a glimpse of the elusive spark that animates our existence.

But amidst all the philosophical pondering and mind-bending questions, one thing remains abundantly clear: the human mind is a marvelously messy masterpiece, full of contradictions, complexities, and infinite possibilities.

Sure, we may never fully understand the mysteries of consciousness or unravel the enigma of free will. But that's okay. Because sometimes, it's the questions themselves that matter most—the act of pondering, exploring, and stretching the boundaries of our understanding.

So, whether you believe in free will, determinism, or just the power of a good existential crisis, one thing's for sure: the mind is a playground of endless wonder, waiting to be explored by intrepid adventurers and curious souls alike.

As we bid adieu to the tangled web of consciousness and free will, let's raise a glass to the sheer audacity of the human spirit—the irrepressible urge to ask questions, seek answers, and dance on the razor's edge between choice and fate.

And who knows? Maybe, just maybe, the next time you find yourself lost in thought, pondering the mysteries of the universe, you'll catch a fleeting glimpse of the truth—the spark of consciousness, the power of choice, the essence of what it means to be human.

But until that day comes, remember this: the journey is what matters most. So keep asking, keep exploring, and keep marveling at the sheer absurdity of it all. After all, the mind is a terrible thing to waste—so why not dive in headfirst and see where it takes you?

Chapter 4: Multiverse Mayhem

In the realm of theoretical physics, the multiverse theory stands as both a captivating concept and a polarizing debate. Imagine the universe as a vast tapestry, each thread representing a different reality, a parallel universe coexisting alongside our own. It's a mind-boggling notion that challenges our understanding of existence and stretches the limits of our imagination.

Picture yourself at the nexus of infinite possibilities, where every decision, every choice, spawns a new universe, branching off into an endless array of potential outcomes. In one universe, you're a world-renowned scientist; in another, you're an astronaut exploring the far reaches of space; and in yet another, you're a humble street performer entertaining passersby with your accordion skills.

Of course, not everyone is sold on the idea of the multiverse. Skeptics argue that it's nothing more than a fanciful hypothesis, devoid of empirical evidence and grounded more in science fiction than scientific fact. According to them, there's no room in the universe for an infinite number of alternate realities— it's simply too big, too complex, to entertain such notions.

But for proponents of the multiverse theory, the allure lies in its potential to explain some of the universe's most perplexing mysteries. From the peculiarities of quantum mechanics to the fundamental constants that govern our reality, the multiverse offers a tantalizing solution—one that suggests that our universe is just one of many, each with its own set of physical laws and properties.

As we delve deeper into the multiverse, we're confronted with questions that push the boundaries of our understanding. What defines a universe? How do they interact with one another? And perhaps most importantly, how do we test such a seemingly untestable theory?

Despite the challenges and controversies surrounding the multiverse, one thing is certain: it's a concept that sparks the imagination and inspires curiosity in equal measure. Whether it's through the lens of theoretical physics or the canvas of science fiction, the multiverse invites us to explore the vast expanse of possibility that lies just beyond the horizon.

So, as we continue our journey through the multiverse, let's keep an open mind and a sense of wonder. After all, in a cosmos as vast and mysterious as ours, who knows what otherworldly marvels await us just beyond the veil of reality?

As we navigate the crazy corridors of the multiverse, we're faced with a humbling truth: our universe is but a tiny speck in the grand tapestry of existence. Whether the multiverse is real or merely a product of our imagination, it challenges us to contemplate the vastness of reality and our place within it.

For some, the multiverse is a source of wonder and inspiration—a testament to the boundless creativity of the cosmos. It invites us to dream big, to imagine worlds beyond our wildest fantasies, and to explore the infinite possibilities that lie just beyond the reach of our understanding.

But for others, the multiverse is a source of skepticism and uncertainty. They argue that it's nothing more than a speculative hypothesis, lacking the empirical evidence needed to support its existence. To them, the multiverse is little more than a flight of fancy—a fascinating idea, perhaps, but one that remains firmly rooted in the realm of theory.

Yet, regardless of where one stands on the matter, there's no denying the allure of the multiverse. It captures our imagination, challenges our assumptions, and invites us to contemplate the mysteries of existence in all their complexity.

So, as we say goodbye to the multiverse and continue our journey through the cosmos, let us carry with us the lessons learned from our exploration. Let us remain open to the wonders of the universe, willing to embrace the unknown and the unexplained.

For in the end, it is our curiosity, our thirst for knowledge, that propels us ever forward, driving us to seek answers to the questions that have plagued humanity since time immemorial. And whether those answers lie within the confines of our own universe or beyond its boundaries, one thing is certain: the quest for understanding is a journey without end, a voyage of discovery that knows no bounds.

So, let us press on, with hearts full of wonder and minds open to the infinite possibilities that lie before us. For in the vast expanse of the multiverse, there are worlds to explore, mysteries to unravel, and wonders beyond imagining.

Chapter 5: The Fermi Paradox

Alright, folks, buckle up and prepare to dive headfirst into the mind-bending world of the Fermi Paradox. This one's a real head-scratcher, so grab your thinking caps and get ready to ponder the ultimate cosmic conundrum: If the universe is teeming with potential life, why haven't we heard from our extraterrestrial neighbors yet?

Picture this: the universe is a vast and seemingly endless expanse, filled with billions upon billions of stars, each with its own retinue of planets. Surely, in such a vast cosmic playground, we can't be the only ones playing the game, right?

That's where the Fermi Paradox comes in. Named after the Italian physicist Enrico Fermi, who famously asked, "Where is everybody?" this paradox asks us to consider the apparent contradiction between the high probability of extraterrestrial life and the lack of any convincing evidence for it.

Now, you might be thinking, "But wait, haven't we been searching for signs of intelligent life for decades?" And you'd be right. From radio telescopes scanning the skies for alien signals to space probes exploring distant worlds for signs of microbial life, we've left no stone unturned in our quest to find our cosmic neighbors.

So why, then, haven't we found any evidence of extraterrestrial civilizations?

It's enough to make even the most hardened skeptic scratch their head in bewilderment.

Of course, there are plenty of theories to go around. Some folks think that we're simply looking in the wrong places, like trying to find a needle in a cosmic haystack. After all, the universe is a big place, and our technology is still in its infancy—maybe we just haven't cast our net wide enough yet.

Others argue that intelligent life is a rare and precious commodity, scattered thinly across the cosmos like grains of sand on a cosmic beach. According to them, we might be the only advanced civilization in our corner of the galaxy, destined to wander the stars alone for eons to come.

But perhaps the most intriguing theory of all is the notion of the "Great Filter." This idea suggests that there's a cosmic obstacle—a barrier of sorts—that prevents intelligent life from emerging and flourishing on a cosmic scale. Whether it's a cataclysmic event like a gamma-ray burst or a technological bottleneck like nuclear war, the Great Filter looms large, casting a shadow of uncertainty over our cosmic aspirations.

So, what's the truth? Well, that's for you to ponder, dear reader. Whether you're a believer in the possibility of extraterrestrial life or a skeptic who thinks we're alone in the universe, one thing's for sure: the Fermi Paradox is a puzzle that's bound to keep us scratching our heads for years to come.

So, as we embark on this journey through the cosmic enigma of the Fermi Paradox, let's keep an open mind and a healthy sense of curiosity. After all, the universe is a big, strange, and wonderfully mysterious place—and who knows what secrets it might still be hiding, just beyond the reach of our understanding?

As we navigate the twists and turns of the Fermi Paradox, it's clear that this cosmic puzzle is more than just an intellectual curiosity—it's a reflection of

CHAPTER 5: THE FERMI PARADOX

our deepest hopes, fears, and aspirations. From the possibility of discovering alien civilizations to the sobering realization that we might be alone in the universe, the Fermi Paradox forces us to confront our place in the cosmos and our place in the grand tapestry of existence.

But amidst the uncertainty and speculation, one thing remains abundantly clear: the search for extraterrestrial life is a journey worth taking. Whether we find evidence of intelligent civilizations beyond our own or whether we continue to explore the cosmos in solitude, the quest for knowledge and understanding is a noble pursuit—one that speaks to the very essence of what it means to be human.

So, as we ponder the mysteries of the Fermi Paradox and wrestle with the cosmic questions it raises, let us remember that the journey is just as important as the destination. Whether we find answers or whether we're left with more questions than we started with, the pursuit of truth is a noble endeavor—one that enriches our lives and expands our horizons in ways we never thought possible.

And who knows? Maybe, just maybe, one day we'll look up at the stars and catch a glimpse of a distant civilization, sending out a beacon of hope across the vast expanse of space. Or maybe we'll find ourselves alone in the darkness, contemplating the wonders of the cosmos in solitude.

But regardless of what the future holds, one thing's for sure: the search for extraterrestrial life will continue to captivate our imagination and inspire us to reach for the stars. So let us press on, with hearts full of wonder and minds open to the infinite possibilities that lie just beyond the horizon.

Chapter 6: The Meaning of Life

Alright, gather 'round for a philosophical rollercoaster ride as we tackle the age-old question: What's the meaning of life? It's the ultimate head-scratcher, the granddaddy of all existential conundrums, and today, we're diving headfirst into the abyss in search of some semblance of an answer.

Now, before we get too lost in the weeds of existential angst, let's take a step back and consider what we're really asking here. When we talk about the meaning of life, are we asking for some grand, cosmic purpose handed down from on high? Or are we simply looking for a reason to get out of bed in the morning and face another day?

It's a tricky question, to be sure. Some folks think that the meaning of life is all about finding happiness and fulfillment—following your passions, chasing your dreams, and living life to the fullest. After all, if life's just a cosmic accident, we might as well make the most of it, right?

Others take a more philosophical approach, arguing that the meaning of life is something deeper, something more profound than mere happiness or success. According to them, it's about finding purpose in the face of adversity, seeking truth in a world of uncertainty, and making a difference in the lives of others.

But here's the kicker: maybe there is no one-size-fits-all answer to the meaning of life. Maybe it's different for everyone, a uniquely personal journey

CHAPTER 6: THE MEANING OF LIFE

of self-discovery and growth. After all, what's meaningful to one person might be utterly meaningless to another—like trying to explain the appeal of pineapple on pizza to someone who's never tried it.

So, where does that leave us? Well, for starters, it leaves us with a whole lot of questions and not a whole lot of answers. But maybe that's okay. Maybe the meaning of life isn't something to be discovered once and for all, but rather something to be explored and pondered, like a well-worn hiking trail through the wilderness of existence.

As we venture forth into the murky waters of existential inquiry, let's remember to approach the question of life's meaning with an open mind and a healthy dose of skepticism. After all, the universe is a big, strange, and wonderfully mysterious place, and trying to boil it down to a single, tidy answer is like trying to fit a square peg into a round hole.

So, grab your philosophical thinking caps and your sense of humor, and let's embark on this journey through the puzzle of existential philosophy. Who knows what we'll find? Maybe some profound insights, maybe a few laughs, or maybe just a newfound appreciation for the sheer absurdity of it all.

But regardless of what awaits us at the end of this philosophical rabbit hole, one thing's for sure: the quest for meaning is a journey worth taking, a journey that enriches our lives and deepens our understanding of the human experience. So let's dive in headfirst and see where it takes us.

As we navigate the swirling currents of existential inquiry, it's easy to feel overwhelmed by the sheer magnitude of the question we're grappling with. What is the meaning of life? It's a query that has plagued humanity since time immemorial, and yet, it remains as elusive as ever.

But amidst the uncertainty and the ambiguity, there's a glimmer of hope—a spark of insight that illuminates the darkness and points us towards a deeper understanding of our existence.

Perhaps the meaning of life isn't something to be found in grandiose gestures or lofty ideals, but rather in the small moments of everyday existence. Maybe it's in the laughter of loved ones, the beauty of a sunset, or the simple joy of sharing a meal with friends.

Or maybe, just maybe, the meaning of life lies in the connections we forge with others, the impact we have on the world around us, and the legacy we leave behind. After all, what's the point of it all if not to make a difference, to leave the world a little better than we found it?

Of course, we may never fully understand the true meaning of life, and that's okay. Because sometimes, it's the journey itself that matters most—the quest for meaning, the search for truth, and the exploration of the human condition.

So, as we ponder the mysteries of existence and grapple with the enigma of life's meaning, let's remember to approach the question with humility, curiosity, and a healthy dose of humor. After all, life is too short to take too seriously, and sometimes, the best way to find meaning is to simply embrace the absurdity of it all.

As we part ways with this chapter on the meaning of life, let's carry with us the lessons learned and the insights gained. And who knows? Maybe, just maybe, the next time we find ourselves pondering the big questions of existence, we'll stumble upon a glimmer of truth, a spark of enlightenment, or perhaps just a good joke to lighten the mood.

But until that day comes, let's continue to explore, to question, and to marvel at the sheer wonder of the human experience. After all, in a universe as vast and mysterious as ours, there's always something new to discover, something new to learn, and something new to ponder.

Chapter 7: Infinity and Finite Minds

Welcome, wanderers of the cosmos, to a chapter that's sure to stretch your mental muscles and tickle your brain cells—the curious case of infinity and its confounding relationship with our finite minds. Get ready to wrap your head around concepts so vast they make the universe itself seem like a mere speck of dust in comparison.

Now, let's start by tackling the big question: what exactly is infinity? Is it just a mathematical abstraction, a convenient way to describe things that go on forever, like the number line stretching out into eternity? Or is it something more profound, a glimpse into the boundless expanse of the universe itself?

Well, strap yourselves in, because we're about to dive headfirst into the infinite abyss and see where it takes us. But fair warning: trying to wrap your mind around infinity is like trying to catch a cloud with your bare hands—fleeting, elusive, and bound to leave you scratching your head in confusion.

Take, for example, the concept of infinity in mathematics. Sure, we can talk about infinite sets, infinite series, and even infinite decimals—but what does it really mean for something to go on forever? It's enough to make even the most seasoned mathematician break out in a cold sweat.

And then there's the philosophical side of infinity—the idea that the universe itself might be infinite, stretching out into eternity with no end in sight. It's a notion that's both exhilarating and terrifying, like staring into the void and realizing just how small and insignificant we really are.

But here's the kicker: our finite minds aren't really equipped to deal with infinity. We're like ants trying to comprehend the vastness of the cosmos, or goldfish trying to grasp the concept of quantum mechanics. It's a futile endeavor, to say the least.

Of course, that hasn't stopped us from trying. From the ancient Greeks pondering the nature of infinity to modern-day physicists grappling with the implications of an infinite universe, humanity has been obsessed with the concept for millennia.

So, what's the takeaway from all this existential pondering? Well, for starters, it's a reminder of just how small we really are in the grand scheme of things. In a universe that's infinite in both space and time, our brief existence on this tiny blue dot we call Earth is little more than a blip on the cosmic radar.

But rather than feeling overwhelmed by our insignificance, let's embrace the wonder of it all. Let's marvel at the sheer audacity of the cosmos, the mind-bending mysteries that lie just beyond the reach of our understanding.

Because in the end, that's what infinity is all about—a reminder of just how vast and wondrous the universe truly is, and a humbling acknowledgment of our place within it. So, as we journey deeper into the realms of infinity and finite minds, let's do so with a sense of awe, curiosity, and just a hint of existential dread. After all, it's a big, strange, and wonderfully mysterious universe out there—so let's dive in and see where infinity takes us.

As we near the end of our expedition into the dizzying realms of infinity and finite minds, it's important to reflect on what we've learned—and perhaps more importantly, what we've yet to understand. Infinity, with its tantalizing allure and perplexing mysteries, remains an enigma that continues to elude our grasp.

But maybe that's the point. Maybe infinity isn't meant to be understood or quantified, but rather appreciated for the profound sense of wonder and awe it inspires within us. Like a vast cosmic tapestry stretching out into eternity, infinity invites us to contemplate the boundless expanse of the universe and our place within it.

As we ponder the implications of infinity and its relationship with our finite minds, let's remember that the journey itself is just as important as the destination. Whether we're grappling with the concept in the realm of mathematics, philosophy, or cosmology, the pursuit of understanding is a noble endeavor—one that enriches our lives and expands our horizons in ways we never thought possible.

So, as we say farewell to the infinite expanse of possibility and return to the familiar shores of finite reality, let's carry with us the lessons learned and the insights gained. And who knows? Maybe, just maybe, the next time we find ourselves staring up at the stars, we'll catch a glimpse of the infinite majesty of the cosmos—and feel a sense of wonderment and awe unlike anything we've ever known.

But until that day comes, let's continue to marvel at the sheer audacity of the universe, the mind-bending mysteries that lie just beyond the reach of our understanding. After all, in a cosmos as vast and wondrous as ours, there's always something new to discover, something new to learn, and something new to ponder.

Chapter 8: The Simulation Hypothesis

Prepare to have your minds blown as we delve into one of the most mind-bending concepts to come out of the digital age—the Simulation Hypothesis. Picture this: What if our reality, the world we perceive around us, is nothing more than a hyper-advanced computer simulation? It's like living in a cosmic video game, only without the respawn button.

Now, before you dismiss this idea as pure science fiction, let's take a moment to consider the evidence—or lack thereof. After all, as we plunge headfirst into the rabbit hole of the Simulation Hypothesis, we're faced with a tantalizing question: What if we're all just ones and zeros in some cosmic code?

But hold onto your hats, because it's about to get even weirder. If our reality is indeed a simulation, then who—or what—is behind the curtain pulling the strings? Are we the unwitting subjects of some hyper-intelligent cosmic programmer, or are we merely characters in a cosmic sandbox, left to our own devices to explore and discover the rules of our simulated universe?

And what about morality, consciousness, and the nature of existence itself? If our reality is nothing more than a computer program running on some cosmic supercomputer, then what does that say about the very fabric of reality? Are our thoughts, our feelings, our very sense of self nothing more than lines of code, programmed into us by some unseen hand?

CHAPTER 8: THE SIMULATION HYPOTHESIS

Of course, the Simulation Hypothesis raises more questions than it answers. But that's what makes it so darn intriguing. From the philosophical implications of living in a simulated reality to the ethical quandaries of playing god with digital beings, the Simulation Hypothesis forces us to confront some uncomfortable truths about the nature of existence.

But fear not, for we're not alone in our existential crisis. From the halls of academia to the virtual realms of cyberspace, thinkers, scientists, and armchair philosophers alike are grappling with the implications of living in a simulated reality. And while we may never know for sure whether our reality is the real deal or just a really elaborate simulation, one thing's for certain: the Simulation Hypothesis is a concept that's here to stay.

So, as we embark on this journey through the virtual realms of the Simulation Hypothesis, let's keep an open mind and a healthy dose of skepticism. After all, in a universe as strange and mysterious as ours, anything is possible—even the idea that we're all just characters in someone else's cosmic video game.

With that, I invite you to buckle up and prepare for a wild ride through the digital cosmos. Whether we're living in a simulation or not, one thing's for sure: the truth is out there, waiting to be discovered. So let's dive in and see where the rabbit hole takes us.

As we reach the end of our exploration into the Simulation Hypothesis, it's clear that this mind-bending concept raises more questions than it answers. Are we really living in a computer simulation, or is this just a fascinating thought experiment that tickles our imaginations? The truth may be stranger than fiction, but it's also shrouded in mystery.

But amidst the uncertainty and speculation, there's something undeniably captivating about the idea of living in a simulated reality. It challenges our assumptions about the nature of existence and forces us to confront the limits of our understanding. Are we the masters of our own destiny, or are we merely pawns in someone else's cosmic game?

Of course, the Simulation Hypothesis isn't without its critics. Skeptics argue that there's simply not enough evidence to support the idea that our reality is nothing more than a computer program. After all, just because something *could* be true doesn't mean that it *is* true.

But perhaps that's missing the point. Maybe the real value of the Simulation Hypothesis lies not in whether it's true or false, but in the questions it raises and the conversations it sparks. From debates about the nature of consciousness to discussions about the ethics of creating digital beings, the Simulation Hypothesis forces us to think critically about our place in the universe.

So, as we say goodbye to the virtual realms of the Simulation Hypothesis, let's carry with us the lessons learned and the insights gained. Whether we're living in a simulation or not, one thing's for certain: the quest for knowledge and understanding is a journey without end.

Chapter 9: The Uncertainty Principle

Get ready to dive into the wonderfully wacky world of quantum mechanics as we tackle one of its most mind-bending concepts—the Uncertainty Principle. Strap in, because things are about to get a little fuzzy.

Now, imagine you're trying to measure the position and momentum of a subatomic particle. Seems simple enough, right? Wrong! According to the Uncertainty Principle, as soon as you try to pin down one of these properties with certainty, the other becomes as elusive as a cat on a hot tin roof.

It's like trying to juggle flaming torches while riding a unicycle on a tightrope—sure, you might be able to pull it off, but chances are you'll end up with singed eyebrows and a bruised ego.

But fear not, for the Uncertainty Principle isn't just about confounding physicists and giving philosophers headaches. It's also a window into the weird and wonderful world of quantum mechanics, where the rules of reality are more like guidelines and the laws of physics play fast and loose with our expectations.

So, what exactly does the Uncertainty Principle tell us about the nature of reality? Well, for starters, it suggests that the universe is a lot fuzzier around the edges than we might like to think. It's like trying to take a picture of a speeding bullet—it's there one moment and gone the next, leaving you scratching your head and wondering what just happened.

But here's where things get really interesting. According to the Uncertainty Principle, the very act of measuring a particle's position or momentum can alter its behavior, like trying to observe a wild animal without disturbing its natural habitat. It's as if the universe itself is playing a game of cosmic hide-and-seek, and we're just along for the ride.

Of course, this raises all sorts of philosophical questions about the nature of reality and our place within it. Are we merely observers in a universe that's fundamentally uncertain, or do we play a more active role in shaping the world around us? It's enough to make your head spin faster than a subatomic particle in a particle accelerator.

But amidst the uncertainty and the confusion, there's a certain beauty in the Uncertainty Principle. It reminds us that the universe is far stranger and more mysterious than we can possibly imagine, and that our understanding of reality is always evolving, always expanding, like a cosmic game of cosmic cat-and-mouse.

So, as we embark on this journey through the quantum weirdness of the Uncertainty Principle, let's do so with a sense of wonder, curiosity, and just a hint of humility. After all, in a universe as strange and wonderful as ours, there's always something new to discover, something new to learn, and something new to ponder.

As we near the end of our journey through the mind-bending maze of the Uncertainty Principle, it's worth taking a moment to reflect on what we've learned—and what we're still trying to wrap our heads around. Quantum mechanics, with its strange and counterintuitive rules, has a way of challenging our most deeply-held beliefs about the nature of reality.

But amidst the confusion and the uncertainty, there's a certain beauty in

CHAPTER 9: THE UNCERTAINTY PRINCIPLE

the quantum weirdness of the Uncertainty Principle. It reminds us that the universe is far more complex and mysterious than we can possibly imagine, and that our understanding of reality is always evolving, always expanding, like a cosmic puzzle with no solution in sight.

So, what's the takeaway from all this quantum strangeness? Well, for starters, it's a reminder that the universe is a lot weirder than we give it credit for. From particles that can be in two places at once to waves that behave like particles, the world of quantum mechanics is a topsy-turvy place where the rules of reality are more like suggestions than laws.

But rather than feeling overwhelmed by the uncertainty of it all, let's embrace the wonder and the awe of the quantum world. Let's marvel at the sheer audacity of the universe, the mind-bending mysteries that lie just beyond the reach of our understanding.

Because in the end, that's what makes the Uncertainty Principle so darn intriguing. It challenges us to think outside the box, to question our assumptions, and to embrace the unknown with open arms.

So, as we depart from the quantum weirdness of the Uncertainty Principle, let's carry with us the lessons learned and the insights gained. And who knows? Maybe, just maybe, the next time we find ourselves grappling with the mysteries of the quantum world, we'll stumble upon a glimmer of truth or a newfound appreciation for the sheer wonder of it all.

But until that day comes, let's continue to explore, to question, and to marvel at the mysteries of the universe. After all, in a cosmos as vast and wondrous as ours, there's always something new to discover, something new to learn, and something new to ponder.

Chapter 10: Existentialism and Absurdism

Buckle up for a wild ride through the existential rabbit hole as we delve into the intriguing philosophies of existentialism and absurdism. Get ready to ponder the big questions of life, the universe, and everything in between—all while trying to maintain your sense of humor.

Now, let's start by unpacking existentialism, shall we? Picture this: you're staring into the void, contemplating the meaning of your existence, and suddenly it hits you like a ton of bricks—you are utterly and completely alone in a cold, indifferent universe. Cheery stuff, right?

But fear not, for existentialism isn't all doom and gloom. At its core, it's a philosophy that celebrates the freedom and individuality of the human spirit, urging us to embrace our existence with passion, authenticity, and a healthy dose of existential dread.

So, what does it mean to live authentically in a world that often feels meaningless and absurd? Well, for starters, it means confronting the harsh realities of existence head-on, rather than burying our heads in the sand and pretending everything's hunky-dory.

But don't worry, existentialism isn't all about staring into the abyss and contemplating the futility of it all. It's also about finding meaning and purpose in the face of adversity, about carving out our own path in a universe that often

CHAPTER 10: EXISTENTIALISM AND ABSURDISM

seems indifferent to our struggles.

Now, if existentialism is the brooding teenager of philosophical movements, then absurdism is its quirky, eccentric cousin who marches to the beat of their own drum. Picture this: you're wandering through the cosmic carnival of existence, and suddenly you realize that the whole thing is utterly, irredeemably absurd. Cue the existential crisis!

But here's the kicker: rather than succumbing to despair or nihilism, absurdism urges us to embrace the absurdity of existence with open arms. It's like laughing in the face of the void, flipping it the bird, and saying, "Bring it on, universe!"

So, what's the takeaway from all this existential pondering and absurd whimsy? Well, for starters, it's a reminder that life is a strange and wonderful adventure, full of twists, turns, and unexpected surprises. Whether we're grappling with the meaning of existence or embracing the absurdity of it all, one thing's for certain: the human spirit is nothing if not resilient.

So, as we dive headfirst into the murky waters of existentialism and absurdism, let's do so with a sense of curiosity, humor, and a healthy dose of skepticism. After all, in a world as bizarre and beautiful as ours, there's always something new to discover, something new to learn, and something new to ponder.

With that, I invite you to join me on this existential odyssey through the realms of philosophy. Whether we're contemplating the meaning of life or laughing in the face of the absurd, one thing's for certain: it's going to be one heck of a ride. So, let's strap in and see where the journey takes us.

As we reach the end of our journey through the enigma of existentialism and absurdism, it's worth taking a moment to reflect on the insights gained and the questions raised. From grappling with the meaning of existence to embracing the absurdity of it all, we've traversed a landscape as varied and complex as the human experience itself.

But amidst the existential angst and the philosophical musings, there's a certain beauty in the chaos of existence. Whether we find meaning in the relationships we form, the passions we pursue, or the simple act of being alive, one thing's for certain: life is a journey worth taking, even if the destination remains uncertain.

So, what's the takeaway from all this philosophical pondering and existential soul-searching? Well, for starters, it's a reminder that the human spirit is nothing if not resilient. In a world that often feels meaningless and absurd, we have the power to create our own meaning, to find purpose in the face of adversity, and to embrace the absurdity of existence with open arms.

But perhaps more importantly, our journey through existentialism and absurdism serves as a reminder of the interconnectedness of all things. Whether we're confronting the void or laughing in its face, we're all in this cosmic carnival together, navigating the twists and turns of existence as best we can.

Chapter 11: The Chicken or the Egg

Alright, folks, prepare to dive headfirst into one of the oldest conundrums known to humanity—the chicken or the egg question. Get ready to untangle the philosophical knot that has puzzled thinkers, scholars, and dinner table debaters for centuries. It's time to crack open this eggcellent puzzle and see what lies inside.

Now, let's start by setting the scene. Picture this: you're sitting at your kitchen table, staring at a plate of scrambled eggs, pondering the age-old question— did the chicken come first, or was it the egg? It's enough to make your head spin faster than a whisk in a mixing bowl.

But fear not, for we're about to embark on a journey through the tangled web of chicken-and-egg logic, armed with nothing but our wits and a healthy appetite for philosophical pondering. So, let's roll up our sleeves, sharpen our minds, and dive into the deliciously perplexing world of the chicken or the egg.

Now, one might think that this question has a simple answer—after all, chickens hatch from eggs, right? But hold onto your frying pans, because things are about to get a little more scrambled than you might expect. You see, the chicken or the egg question isn't just about biology; it's also about the nature of causality, the origins of life, and the mysteries of existence itself.

So, where do we begin? Well, let's start by examining the chicken's side of the argument. According to chicken enthusiasts everywhere, the chicken must have come first—after all, how else could there be eggs without chickens to lay them? It's a compelling argument, to be sure, but is it enough to crack the case wide open?

On the other hand, egg aficionados argue that the egg must have come first—after all, how else could there be chickens without eggs to hatch them? It's a classic chicken-and-egg paradox that has stumped philosophers for centuries, and it shows no signs of being cracked anytime soon.

But wait, there's more! Some enterprising thinkers have even proposed alternative explanations for the origins of chickens and eggs, ranging from the divine creation of the universe to the bizarre realms of quantum mechanics. It's enough to make your brain scramble faster than a dozen eggs on a hot skillet.

With that said, I invite you to join me on this eggciting adventure as we crack open the chicken or the egg question and see what lies inside. Whether we're pondering the origins of life or simply enjoying a delicious omelet, one thing's for certain: the chicken or the egg question is a puzzle worth pondering. So, let's dig in and see where this eggcellent adventure takes us!

As we near the end of our journey through the delightful maze of the chicken or the egg question, it's worth taking a moment to reflect on the insights gained and the yolks cracked along the way. From pondering the origins of life to contemplating the mysteries of existence, we've traversed a landscape as vast and varied as the breakfast buffet at a five-star hotel.

But amidst the philosophical musings and the egg-related puns, there's a certain joy in the quest for knowledge, in the pursuit of understanding, and in the simple act of pondering life's most perplexing questions. Whether we're arguing the merits of chickens or eggs or simply enjoying the delicious

CHAPTER 11: THE CHICKEN OR THE EGG

ambiguity of it all, one thing's for certain: the journey is its own reward.

So, what's the verdict? Well, as much as we might like to have a definitive answer to the chicken or the egg question, the truth is that it's likely to remain an unsolvable mystery for the ages. After all, when it comes to the mysteries of the universe, sometimes it's best to embrace the uncertainty and enjoy the ride.

But fear not, for the journey doesn't end here. Whether we're pondering the origins of life, the nature of existence, or the meaning of it all, there's always another question waiting to be asked, another puzzle waiting to be solved.

Conclusion:

As we reach the end of our journey through the cosmos of deep questions and profound theories, let us pause to reflect on the wonders we've encountered and the insights we've gained. From the timeless enigma of the chicken or the egg to the mind-bending mysteries of quantum mechanics and the existential ponderings of human existence, we've traversed a landscape as vast and varied as the universe itself.

But amidst the uncertainty and the ambiguity, there's a certain beauty in the quest for knowledge, in the pursuit of understanding, and in the simple act of pondering life's most profound questions. Whether we've found answers or merely stumbled upon more questions, one thing's for certain: the journey has been its own reward.

So, as we say goodbye to the realms of thought and inquiry, let us carry with us the lessons learned and the insights gained. Whether we find ourselves contemplating the mysteries of existence or simply marveling at the sheer wonder of it all, let us never lose sight of the awe and the curiosity that drive us forward.

And though our journey may be ending, let us remember that the quest for knowledge knows no bounds. There are always more questions to ask, more mysteries to unravel, and more wonders to behold. So let us continue to explore, to question, and to ponder, for the universe is a vast and wondrous

CONCLUSION:

place, filled with infinite possibilities.

With that, I bid you farewell, fellow travelers of the cosmos. May your journey through the depths of thought and inquiry be filled with wonder, curiosity, and a sense of awe at the majesty of the universe. Until we meet again, happy pondering!

Printed in Dunstable, United Kingdom